# GLASS
# AND THE ENVIRONMENT

## SUSAN CACKETT

**Stargazer Books**

© Aladdin Books Ltd 2005

New edition published in the United States in 2005 by:
Stargazer Books
c/o The Creative Company
123 South Broad Street
P.O. Box 227
Mankato, Minnesota 56002

Printed in UAE

Editor: Katie Harker

Designer:
Pete Bennett – PBD

Illustrator: Louise Nevett

Picture Researcher:
Brian Hunter Smart

Library of Congress Cataloging-in-Publication Data

Cackett, Susan.
  Glass and the environment / by Susan Cackett ; [illustrator, Louise Nevett].–New ed.
   p. cm.– (Resources)
  Includes index.
  ISBN 1-932799-37-0 (alk. paper)
   1. Glass–Juvenile literature.
   2. Glass manufacture–Juvenile literature.
   [1. Glass. 2. Glass manufacture.] I. Nevett, Louise, ill.
   II. Title. III. Resources (North Mankato, Minn.)
TP857.3.C32 2004
666'.1–DC22                                    2003070752

# CONTENTS

# WHAT IS GLASS?

Glass is a strong, transparent material that has been used for centuries to make many familiar objects. Windows, bottles, and drinking glasses are just a few examples. Glass is cheap to produce because it is mostly made from sand.

When hot glass cools, it hardens before the molecules can arrange themselves in a regular way (the way they do in metal, for example). Although glass is rigid and solid at room temperature, scientists sometimes describe glass as a liquid because of the behavior of its molecules. Glass even acts like a liquid. Light passes through it, and when glass is heated it flows easily and can be molded into different shapes.

The glass in a greenhouse traps light and heat. Plants grow well in these conditions.

Ultraviolet light rays reflected

Light rays

White and infrared light rays pass through.

Glass is transparent because of the "loose" arrangement of its molecules. Ultraviolet light rays are reflected, but white and infrared light rays pass through it.

Glass is perfect for making windows because it protects you from the wind, rain, and cold, but still lets the sunlight in.

# WHAT IS GLASS MADE FROM?

Glass is mainly made from the chemical silica (silicon dioxide), which comes from sand. A very high temperature is needed to melt silica, so soda (sodium carbonate) is added to lower the melting point. Silica and soda produce a glass that dissolves in water. This is not suitable for making windows, so limestone (calcium carbonate) is added to make normal, strong glass.

The ingredients can be varied to make special kinds of glass. Adding lead oxide instead of most of the limestone gives a heavy glass that is used to make wine glasses.

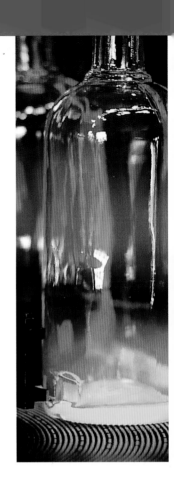

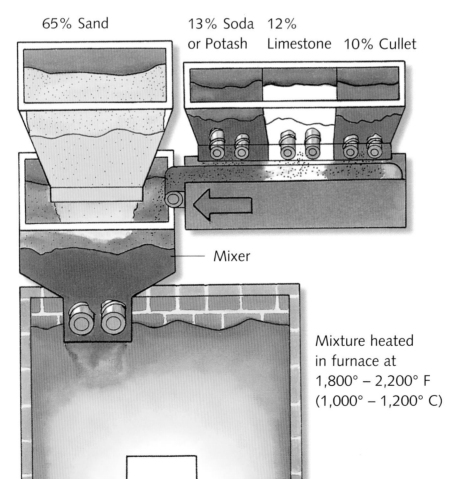

65% Sand    13% Soda   12%
            or Potash  Limestone   10% Cullet

Mixer

Mixture heated
in furnace at
1,800° – 2,200° F
(1,000° – 1,200° C)

## Making glass

The raw materials are mixed together in the right quantities and melted in a huge furnace. The size of the furnace depends on how much glass needs to be made. A typical furnace for flat glass may hold 2,000 tons of molten glass. Some waste glass (cullet) of the same color and type is usually added. Colored glass is made by adding different metal compounds. For example, copper oxide produces blue glass, while chromium compounds give green or yellow glass.

The silica found in sand is the main ingredient used to make glass.

Limestone is added to silica and soda to make glass stronger.

# GLASS FOR WINDOWS

Glass is most commonly used for making windows. Today, windows are made using the "float glass" process, a technique invented in the 1950s.

Earlier methods of glassmaking produced sheets of rough glass that had to be ground and polished to make them smooth enough for windows. This was expensive because a lot of energy was needed. In the float process, the resulting glass is perfectly smooth. This is because the glass floats on a bath of molten metal that is smooth and shiny like a mirror. The bath is surrounded by an inactive gas so there is also nothing to spoil the top surface of the glass.

## Float glass process

The raw materials are melted in a furnace. A ribbon of molten glass then goes into the float bath where it floats on the surface of molten tin. The thickness of the glass can be varied by controlling the rate at which it flows through the float bath.

If glass cools too quickly, it becomes brittle and is no good for normal use. It must therefore be reheated (but not so much as to change its shape) and cooled slowly. This process is known as "annealing." Annealing takes place in a long tunnel called a lehr.

Melting furnace                    Bath of molten tin

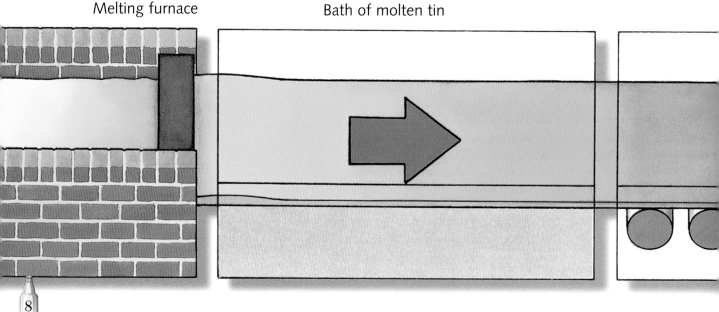

The float glass process has revolutionized the manufacturing of glass windows.

Computer control room

From the annealing lehr, the glass is first washed and then cut up into huge sheets that are lifted off by machine. Any waste glass is collected to be used again. The cutting process and the moving and stacking of the glass are all controlled by computer.

Taken to warehouse

Annealing lehr

Cutting process

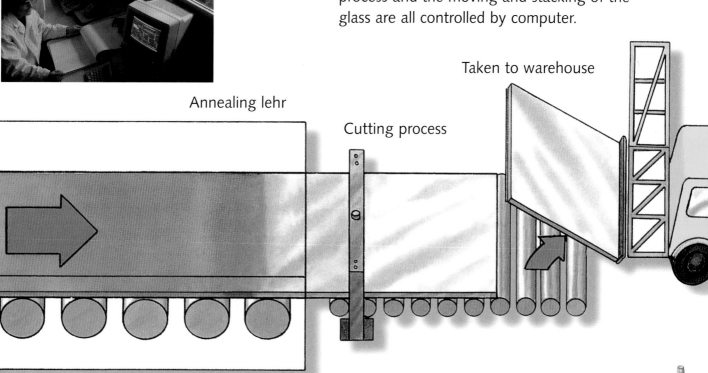

# DIFFERENT KINDS OF WINDOWS

It can get very hot inside a room with big glass windows. This is an advantage for a farmer growing vegetables in a greenhouse, but it can be uncomfortable in an office. Double glazing or special solar control glass is used to help reduce the amount of heat that enters a building.

If the lighting in a room is right, coated glass can be used as a one-way mirror for security purposes. Those on the inside can see out perfectly, but those outside will just see their own reflection. Patterned glass is also used for privacy, in a bathroom window, for example.

To make patterned glass, an old method of glassmaking is used, in which the glass from the furnace is poured between metal rollers to flatten it into a sheet. The pattern is imprinted into the hot glass from the top roller.

Many different designs are possible, some quite complicated. Notice how the pattern makes it more difficult to see through the glass and that it is smooth on one side. Patterned glass is often colored for use in decorative screens.

Melting furnace

Annealing

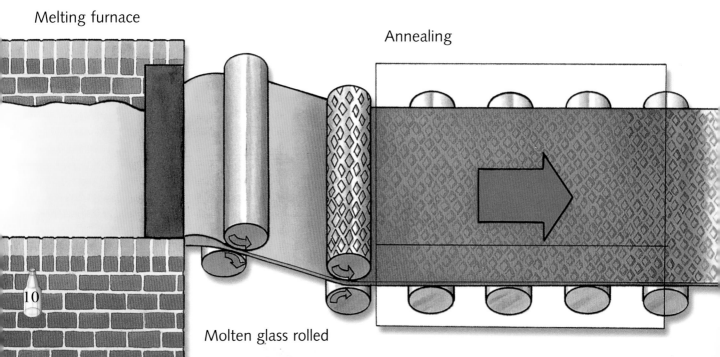

Molten glass rolled

Many offices use coated-glass windows for security purposes.

Double glazed windows provide better insulation than normal windows. They are made from two panes of glass with a layer of air trapped between them. In winter, double glazing reduces the amount of heat that escapes and keeps the inside warm. In summer, it limits the amount of sunlight coming in, preventing a room from becoming too warm. Double glazing also reduces noise pollution in a similar way.

In office blocks, where the whole wall may be made of glass, solar control glass is often used to help keep the heat out, while still letting the light in. The glass may be colored or coated with a thin layer of metal.

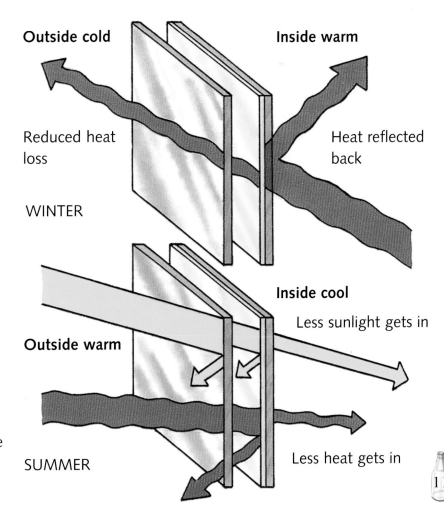

**Outside cold**  **Inside warm**

Reduced heat loss

Heat reflected back

WINTER

**Outside warm**

**Inside cool**

Less sunlight gets in

SUMMER

Less heat gets in

11

# BOTTLES AND JARS

Bottles and jars are made by blowing air into glass. Today, bottles and jars are mass-produced using machinery. To make a jar, molten glass is dropped into a mold that has a plunger attached to it. The plunger is used to press the rough shape of the jar. The jar is then finished in a second mold by blowing. Bottles cannot be shaped using a plunger because they have a narrow neck. Instead, the glass is blown at both stages.

A big, modern bottle machine has a number of molds operating at the same time. Some factories can make up to 12,000 bottles an hour. As with all glass, bottles must be annealed before they can be used.

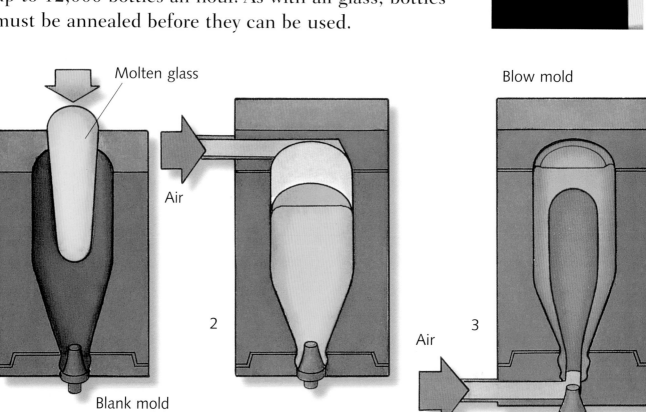

## Making a bottle

A lump of molten glass, known as a "gob," drops into the first mold which is upside down (1). The bottom of the mold is the exact size of the neck of the bottle with a plug in the center. Air is blown in at the top so that the glass is pushed down into the mold to form the neck (2). The plug at the bottom is removed.

Glass bottles are now mass-produced in factories, where they are quicker and cheaper to make.

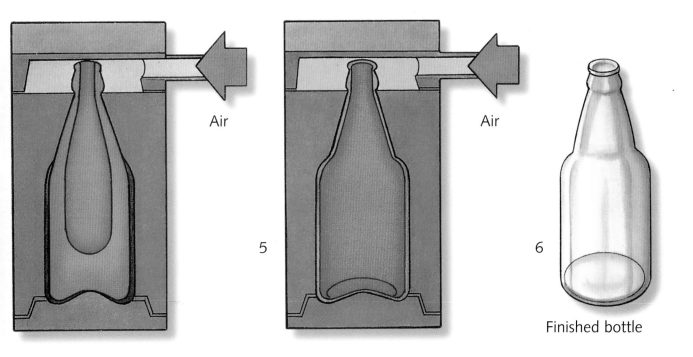

Air

5

Air

6

Finished bottle

Next, a plate is fixed over the top of the mold. More air is blown in from the bottom to form the rough shape, or "parison" (3). The parison is removed, turned over, and placed in the second mold (4), where the final shape is blown (5). The bottle is now ready for annealing (6). When the bottle has cooled, it is stacked and stored, ready to be filled.

# DESIGN AND PRODUCTION

Today, bottles are designed with the aid of a computer. Glass manufacturers discuss the size and shape of a bottle with their customers before it is made. Manufacturers can also work out the minimum amount of glass that is needed to make a bottle.

Modern design has led to bottles that are lighter and cheaper than they used to be, but still strong enough not to break when they are opened or handled. Making sure that a bottle stands up and pours well is also important!

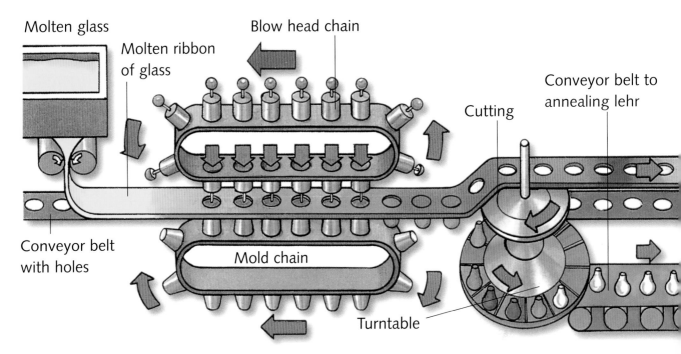

Molten glass

Molten ribbon of glass

Blow head chain

Conveyor belt to annealing lehr

Cutting

Conveyor belt with holes

Mold chain

Turntable

## Making a light bulb

The glass parts of light bulbs, known as "envelopes," are made on a fully automated machine running at high speed. A ribbon of molten glass from the furnace is blown into little blisters by a continuously moving chain of blow heads. Each blister drops into a mold as it passes under the glass ribbon. The glass bulbs are then cut from the ribbon by a rotating disk and collected on a conveyor belt, where they are taken to be annealed.

14

Glass bottles are made in a variety of shapes and sizes.

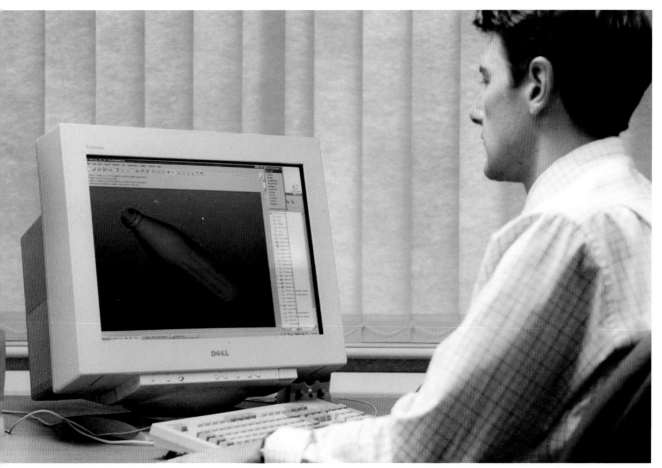

New bottle shapes are designed with the aid of computer images.

# PRESSING AND DRAWING

Glass objects can also be made by pressing or drawing. "Pressing" can be used to make items, like vases, with quite complicated patterns on them. The patterns come from a mold. Molds can be expensive to make, but are stored in the factory to be used again.

"Drawing" is the method used to make tubes like fluorescent lights or scientific glassware. Drawing uses a hollow, rotating tool called a "mandrel." Molten glass is drawn over the mandrel while air is pumped in at the top. The thickness of the tube depends on the pressure of the air and the speed of the drawing.

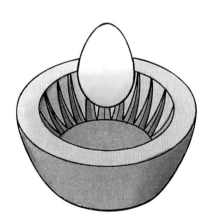

Gob of glass
dropped into mold

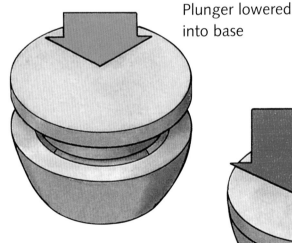

Plunger lowered
into base

Pressure forms
final shape

In pressing, the gob of glass is dropped into an open mold, and the top half of the mold, called the plunger, is pressed down into it. Only dishes and objects that are wide at the top can be made by pressing. It must be possible to get the upper part of the mold in and out. Pressing can be done automatically or by hand.

Finished bowl

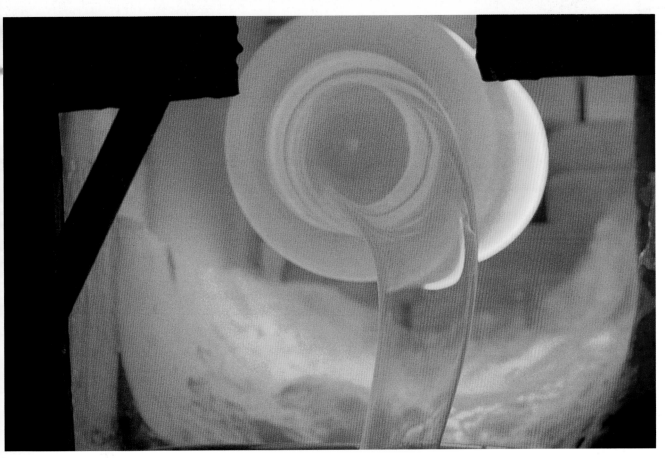

Glass tubing is made with a tool called a mandrel.

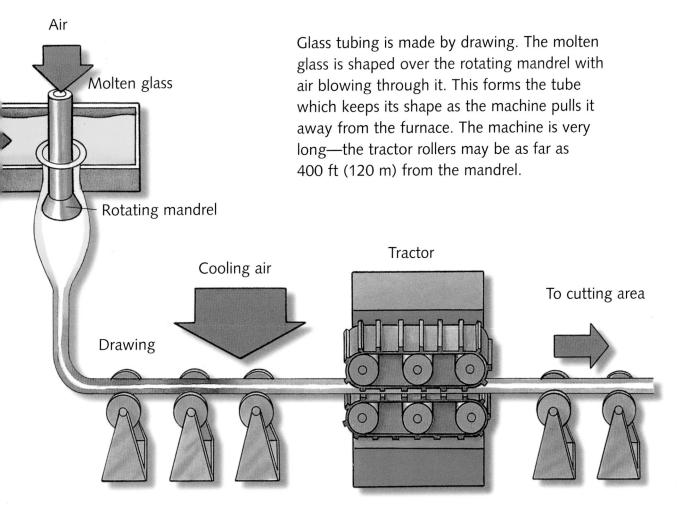

Glass tubing is made by drawing. The molten glass is shaped over the rotating mandrel with air blowing through it. This forms the tube which keeps its shape as the machine pulls it away from the furnace. The machine is very long—the tractor rollers may be as far as 400 ft (120 m) from the mandrel.

Air

Molten glass

Rotating mandrel

Cooling air

Tractor

To cutting area

Drawing

# GLASS FIBER

If molten glass is forced through a small hole, it can be drawn into very fine fibers. This is commonly known as "fiberglass." Short strands are made into thick mats that are used for insulation. Many houses have a layer of fiberglass insulation in their roofs and outer walls to keep the heat in during the winter and out during the summer.

Long strands of glass fiber are also used to reinforce (strengthen) materials such as plastic. The glass fibers make glass-reinforced plastic (GRP), a strong and stiff material that is also light and waterproof. GRP can be brightly colored and is easy to mold into shapes such as "hard-hat" helmets and car bodies.

Molten glass

Spinner

Binding spray

**Making insulation**
To make glass fiber for insulation, the fibers are chopped up and showered onto a moving web.

Curing oven

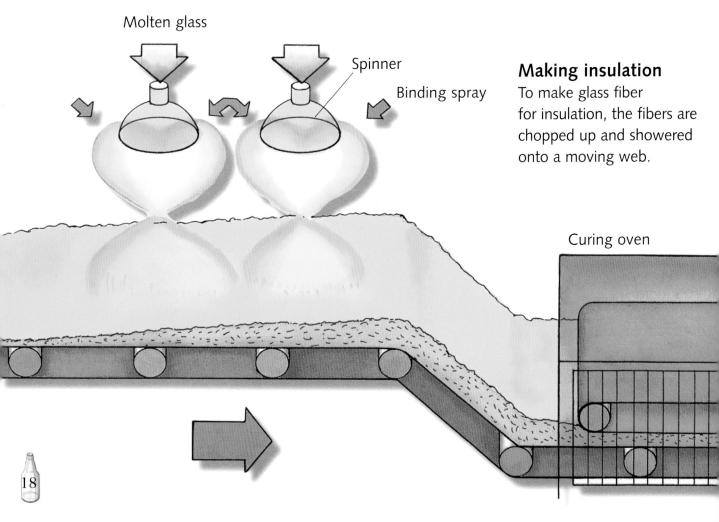

Glass fiber is ideal for making the body of vehicles because it is strong and light.

A special binding spray (glue) is added and heated in an oven to stick the fibers together. The mat is then cut up into convenient sizes for roof, wall, or floor insulation, for example.

Rolling machine

Guillotine

Trimmer    Slitters

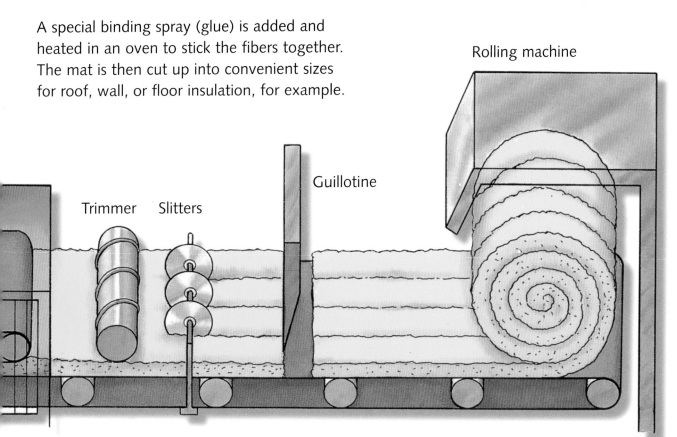

# STRONG GLASS

Glass is strong enough for the normal uses of windows, jars, and bottles in normal use, but it can be broken very easily. When safety is important, glass that has been specially strengthened is used. In the past, cars were fitted with a toughened (heat-treated) glass windshield. When toughened glass breaks, it shatters into small pieces instead of leaving sharp splinters.

Windshields are now made from laminated glass—a glass "sandwich" with a layer of plastic in the middle. Laminated windshields may crack but they don't shatter. This greatly reduces the risk of injury. Other kinds of strong glass include wired glass and bulletproof glass, which is made from several sheets of glass separated by plastic layers.

Laminated and wired glass are both made by rolling. The plastic or wire is sandwiched between two sheets of glass. The strong glass is then annealed and cut to size. Laminated glass is perfectly clear and is ideal for car windshields or store windows, to prevent people from getting hurt if the glass breaks. Wired glass is used where falling glass would be dangerous. If the glass melts in a fire, the pieces are held in place by the wire. Invented in 1895, wired glass was the first safety glass ever made.

Glass

Wire

Glass

WIRED GLASS

Annealing

Glass

Glass

Annealing

Transparent plastic

LAMINATED GLASS

20

Today's windshields are made from laminated glass. On impact, the windshield may crack, but because of its triple layer construction, it will not shatter. This reduces the chances of people being cut or injured by flying glass.

Fighter planes are fitted with a bulletproof windshield—often up to 5 in (12 cm) thick.

# SPECIAL GLASS

Special kinds of glass can be made if other chemicals, such as metal oxides, are added. Borosilicate glass is made from the chemicals silica and boric oxide, and is usually known by the trade name "Pyrex." Pyrex is used to make casserole dishes, chemical glassware, and industrial flasks and pipes, because it does not crack when heated.

Metal oxides, such as zinc, lead, and magnesium oxide, are added to make high quality optical glass for the lenses in cameras, microscopes, telescopes, and some eye glasses. Optical glass is hard to form and expensive to produce. It must be completely transparent so that light passes directly through it, without distortion.

Binoculars and microscopes use optical glass to refract light and make objects look bigger.

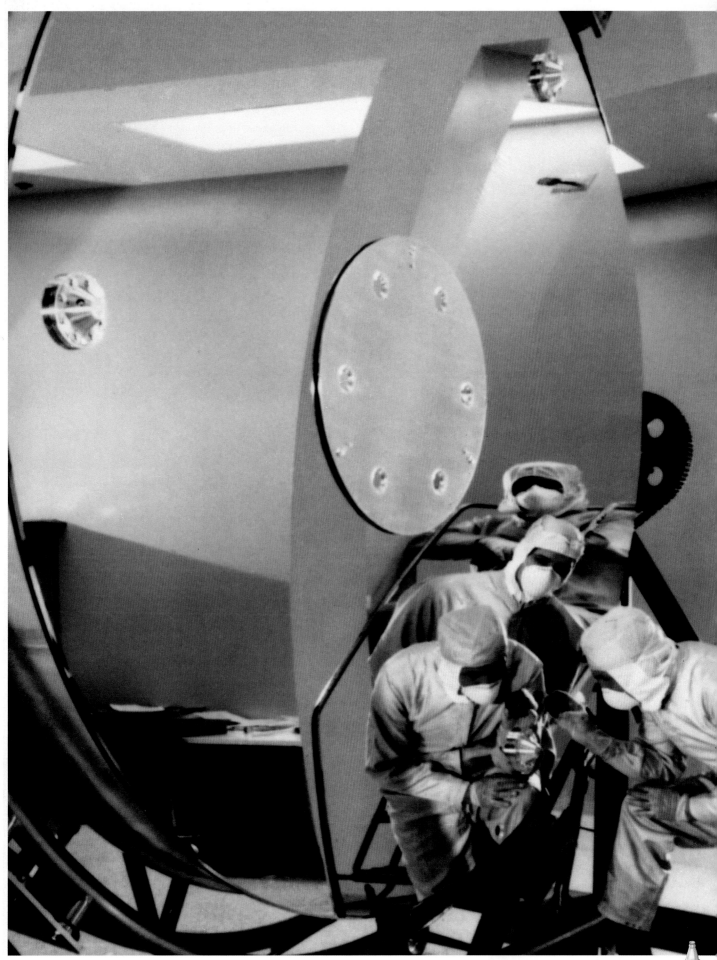

The Hubble space telescope mirrors are precisely shaped and extremely smooth.

# DECORATIVE GLASS

When glass is heated, its surface and shape can be altered in many different ways. Techniques for shaping and coloring glass have been practiced for hundreds of years. Stained glass was first used by wealthy ancient Romans to decorate their villas and palaces. Over the years, clerics began to use stained and tinted glass windows in churches to keep these places of worship cool and dim.

At the end of the 17th century, the Bohemians discovered that adding chalk to glass created a much more brilliant version that, once cooled, was thick enough to engrave with elaborate patterns. Today, decorative glass is still popular and regarded as a highly-skilled craft.

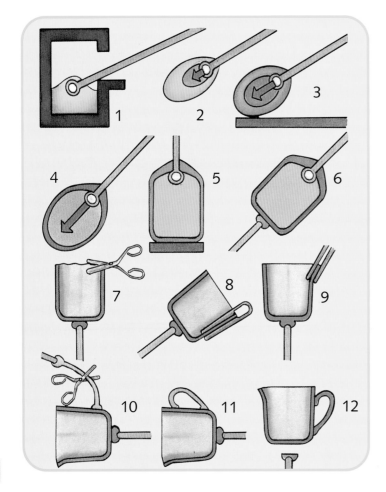

## Handmade glass

In small glassworks and studios, glass is still made by hand. The raw materials are melted in a small furnace. The glassblower gathers a gob of glass onto the end of a long iron pipe, known as a "blowing iron" (1). By blowing down the pipe (2) and shaping the glass against a stone slab (3-5), the glass can be made into the shape required. A solid rod is attached to help hold the article (6), and the top is cut off with shears (7). The glassblower will then finish shaping the work (8-12). If a handle is required, the glassblower will get more glass from the furnace (10-11). Finally, the rod is cut off (12).

## Stained glass

Stained glass windows have been a feature of church architecture for centuries. The design is laid out on a table and small panes of colored glass are mounted in lead frames to make an elaborate picture. The colors are usually produced by adding metal oxides when the glass is made, although extra details may sometimes be painted on.

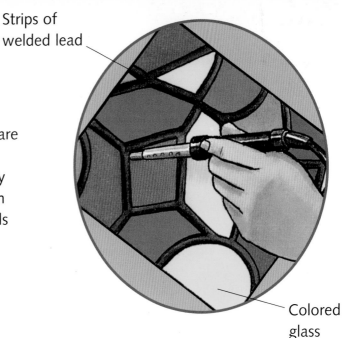

Strips of welded lead

Colored glass

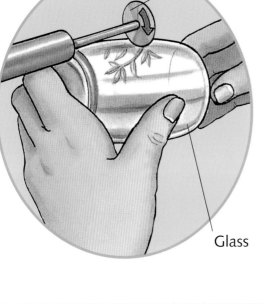

Spinning copper wheel

Glass

## Decorating glass

Handmade glass for wine glasses is often decorated. Lead oxide glass is especially suitable as it sparkles in the light. Deep patterns are made by cutting. More delicate designs are made by engraving using a copper wheel or a high-speed drill, similar to those used by dentists. Glass can also be decorated by "sand blasting," using a gun that fires sand particles.

Decorative glass is still made by hand in many parts of the world.

# THE ENVIRONMENT

Glass is a part of our everyday life, but the processes used to produce this valuable material can also have a damaging effect on our environment. We have come a long way in reducing the environmental impact of glass, with an established recycling industry and more efficient methods of production. However, we still need to do a lot more to reduce the impact that the glass industry has on our planet.

## ◀ Pollution in production

The fuel and raw materials used in glass production release hazardous chemicals, such as sulfur and nitrogen oxides, into the atmosphere. Pollutants can also spread to nearby water sources. Factories now use filters to reduce air pollutants and closely monitor their drainage systems to minimize water pollution.

## ▶ Energy

Extremely high temperatures are needed to melt the raw materials that make glass. Where possible, factories now use electricity as a heat source, instead of valuable natural resources such as gas and oil. Efficient furnace design also helps to prevent heat loss. Using waste glass alongside raw materials enables manufacturers to use lower temperatures, saving energy and reducing the level of oxides that are released into the atmosphere.

### ◄ Tourism and pollution

In the developed world, over 900 lbs (400 kg) of domestic waste (per person) is generated each year. Beautiful locations are popular tourist attractions, but they can easily be spoiled by careless waste disposal. More importantly, some litter can be very dangerous to humans, animals, and other wildlife. Glass bottles left lying around can trap small animals, and broken glass is a serious hazard. Never leave your trash lying around in public places, and if glass gets broken, wrap it up before you dispose of it in a trash can.

### ► Raw materials

Hundreds of thousands of tons of raw materials are quarried each year for the purpose of making glass. Much of this glass is later thrown away as garbage. Meanwhile, the quarrying continues, resulting in scarring of the landscape and the loss of many natural habitats.

### ◄ What you can do

Did you know that recycling glass is one important way that you can help to improve the environment? Recycled glass can be used in glass furnaces to save on raw materials, cut energy costs, and reduce pollutants. If you throw glass away, it will be discarded in landfill sites, wasting precious natural resources. Instead, recycling saves hundreds of thousands of tons of raw materials from being quarried each year and conserves the countryside for everyone.

# GLASS TECHNOLOGY

Glass has been used for many centuries for windows, optical lenses, and decorative purposes. But there are also many technological advances that have been made possible thanks to the unusual properties of glass. Here are just a few examples.

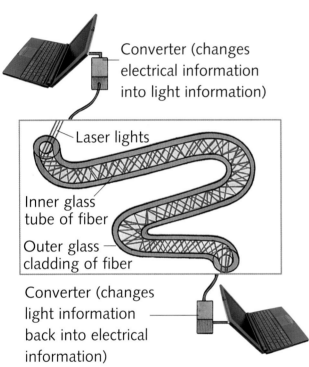

Converter (changes electrical information into light information)

Laser lights

Inner glass tube of fiber

Outer glass cladding of fiber

Converter (changes light information back into electrical information)

## Fiber optics

Thin strands of optically pure glass, as thin as human hair, are used to carry digital information over long distances. Signals are sent along the core of each glass fiber as pulses of laser light and thousands of these fibers are bundled together to form a cable. Fiber optics are light, flexible, and relatively inexpensive. They are ideal for investigative medical instruments used to see inside a patient, and for telephone, television, and computer cables. Unlike metal cables that conduct heat or electricity, fiber-optic signals are not affected by other fibers in the same cable. This means that you can get a clear telephone and television signal at the same time.

## Mirrors

Shiny, smooth surfaces, such as metals, are the best reflectors of light. A mirror, made from a sheet of glass with a thin layer of silver on the back, reflects light almost perfectly. Glass for mirrors must be completely flat so that the image is not distorted. Float glass is ideal. The glass is first washed and then coated with a tin compound. This ensures that the silver deposit is embedded in the surface of the glass. The silver is deposited by the action of several chemicals. It is then covered with copper, red paint, and varnish to protect the layers of metal.

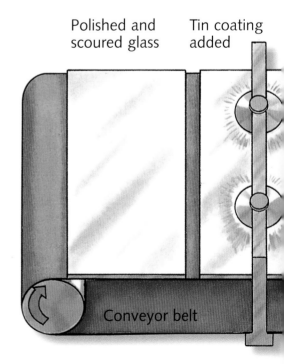

Polished and scoured glass

Tin coating added

Conveyor belt

## Glass ceramics

Glass can be made stronger if its molecules are forced into a regular pattern. Chemical substances are added to glass, and through intense heat treatment, these particles act as "seeds" around which crystals form. The crystallized glass is called glass ceramic. Glass ceramics can be heated or cooled without cracking, so they are ideal for ovens, freezers, stoves, and fireplaces. Glass ceramics are also used for missile and rocket nose cones and as thermal insulation to protect space shuttles as they reenter the earth's atmosphere.

A rocket nose made of glass ceramics

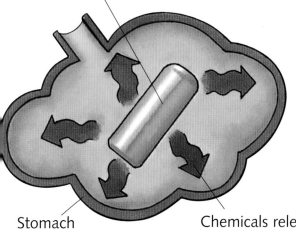

Soluble glass pill containing chemicals

Stomach

Chemicals released into stomach over long period as glass dissolves

## Soluble glass

Glass made from silica and soda, which dissolves in water, has some unusual medical uses. Soluble glass pills can be made containing drugs or vitamins in the center and are particularly useful in veterinary medicine. If the pill is fed to a sheep, for example, the glass slowly dissolves, releasing drugs or vitamins into the stomach. In this way, large doses of medication can enter an animal's bloodstream.

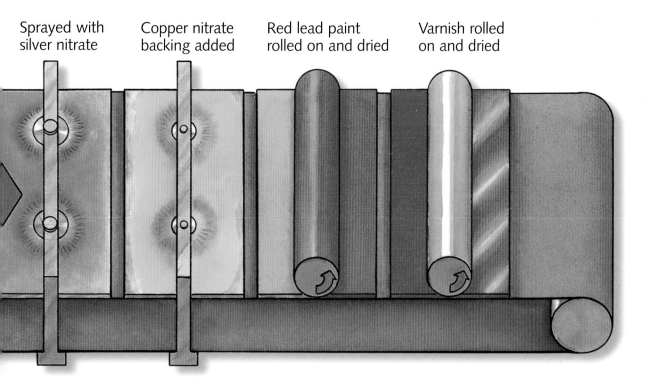

Sprayed with silver nitrate

Copper nitrate backing added

Red lead paint rolled on and dried

Varnish rolled on and dried

# GLASS: THE FUTURE

We have come a long way in our knowledge and understanding of the uses and properties of glass. But glass technology is continually progressing. The table below outlines some of the exciting developments in medicine, science, manufacturing, and architecture that originate from the unusual properties of glass.

| Subject area | Application |
| --- | --- |
| MEDICAL SCIENCE | Bioglass is a manmade material with a chemical composition similar to that of bone. Bioglass is used to bind bone implants to existing human tissue and gradually dissolves after the bone has healed. The ability of bioglass to accelerate the healing process in bone fractures and transplants is a major breakthrough in medical science. |
| DENTISTRY | Bioglass has been found to have a fluoridelike component and scientists believe that it could one day replace fluoride in toothpaste. In the meantime, bioglass is used for root implants and to prevent jawbone deterioration in denture wearers. The implants bond with living tissue and the jawbone gradually reforms on their surface. |
| MICROBIOLOGY | The antibacterial effects of bioglass have led scientists to believe that the material has the capability of protecting plants from harmful insects. If this is true, bioglass-based pesticides could become a common aid used by farmers to combat persistent crop-eating insects. |
| INTERIOR DESIGN | Architects have been developing laminated glass to provide acoustic barriers, UV protection, and solar controlled panels for offices and homes. Their work provides a practical alternative to conventional drapes and blinds. In the future, office partitions, with options of clear or frosted glass, may be activated at the flick of a switch. |
| GLASS PRODUCTION | The glass manufacturing industry is constantly looking to improve energy efficiency and prevent pollution. Reducing the amount of air entering a furnace reduces nitrogen oxide emissions. New and improved materials and coatings also improve energy efficiency and reduce emissions. |

# GLOSSARY

**Annealing**
The reheating and slow cooling of glass to release any internal stress.

**Bioglass**
A special type of glass that has a chemical composition similar to that of bone.

**Cullet**
Waste glass that is recycled.

**Double glazing**
Glass windows made from two panes of glass with a layer of air trapped between them.

**Drawing**
A technique used for making glass tubes. A mandrel is used to draw glass.

**Fiberglass**
The common name for matted fine glass fibers, used as insulation in buildings and in fireproof fabrics.

**Fiber optics**
Cables, made of thin strands of optical glass, used to carry digital information over long distances.

**Furnace**
A large structure lined with bricks where raw materials are melted to make resources, such as glass.

**Glass ceramic**
Crystallized glass, made using an intense heat treatment.

**Gob**
A lump of molten glass.

**GRP**
Glass-reinforced plastic. A type of plastic strengthened with long strands of glass fiber.

**Laminated glass**
Strengthened glass made from two layers of glass with a plastic layer in the middle.

**Lehr**
The part of a glassmaking factory where annealing takes place.

**Mandrel**
A circular core over which glass can be shaped. A mandrel is used to make glass tubing.

**Melting point**
The temperature at which a solid changes state from solid to liquid.

**Molecule**
The simplest unit of a chemical that has two or more atoms held together by chemical bonds.

**Molten glass**
The name for glass before it has hardened into a solid.

**Optical glass**
A special type of glass that contains metal oxides. Optical glass is used for the lenses in cameras, microscopes, telescopes, and some eye glasses.

**Parison**
The partly-formed shape of a bottle or jar.

**Pressing**
A technique used to make glass objects, such as vases. A mold is used for pressing.

**Pyrex**
The trade name for a special type of glass that contains metal oxides. Pyrex does not crack when heated.

**Silica**
A chemical compound of silicon and oxygen, very commonly found in nature as quartz in sand.

**Soluble glass**
A special type of glass that dissolves in water.

**Toughened glass**
Glass that has been strengthened by heating.

# INDEX